All about the Sky

MOON PHASES

Douglas Hicton

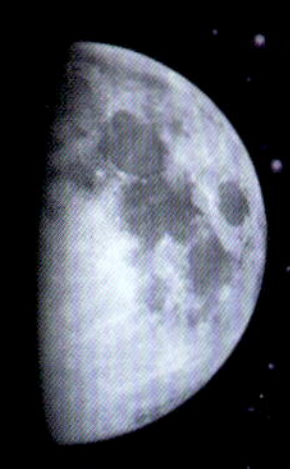

AV2

www.openlightbox.com

Step 1
Go to **www.openlightbox.com**

Step 2
Enter this unique code
BCLRHLGJU

Step 3
Explore your interactive eBook!

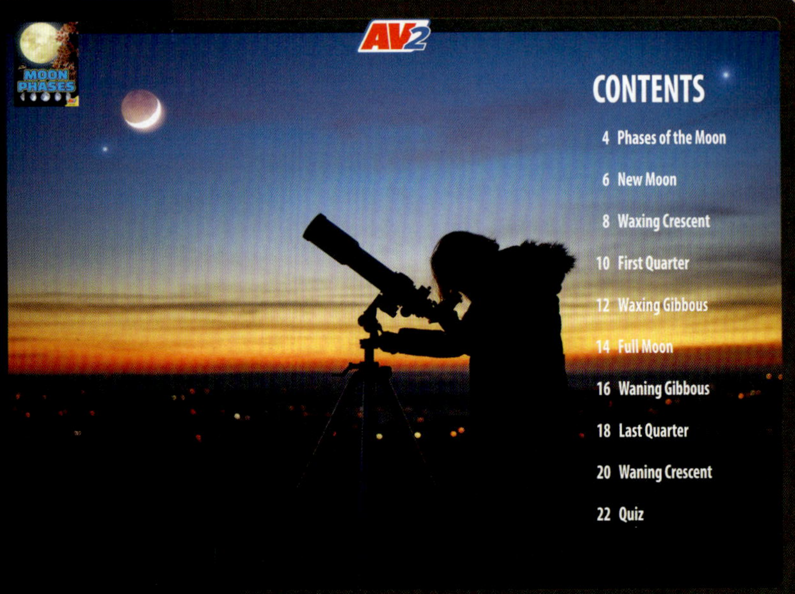

AV2 is optimized for use on any device

Your interactive eBook comes with...

Contents
Browse a live contents page to easily navigate through resources

Audio
Listen to sections of the book read aloud

Videos
Watch informative video clips

Weblinks
Gain additional information for research

Slideshows
View images and captions

Try This!
Complete activities and hands-on experiments

Key Words
Study vocabulary, and complete a matching word activity

Quizzes
Test your knowledge

Share
Share titles within your Learning Management System (LMS) or Library Circulation System

Citation
Create bibliographical references following the Chicago Manual of Style

This title is part of our AV2 digital subscription

1-Year 3–8 Subscription
ISBN 978-1-7911-3306-1

Access hundreds of AV2 titles with our digital subscription.
Sign up for a FREE trial at **www.openlightbox.com/trial**

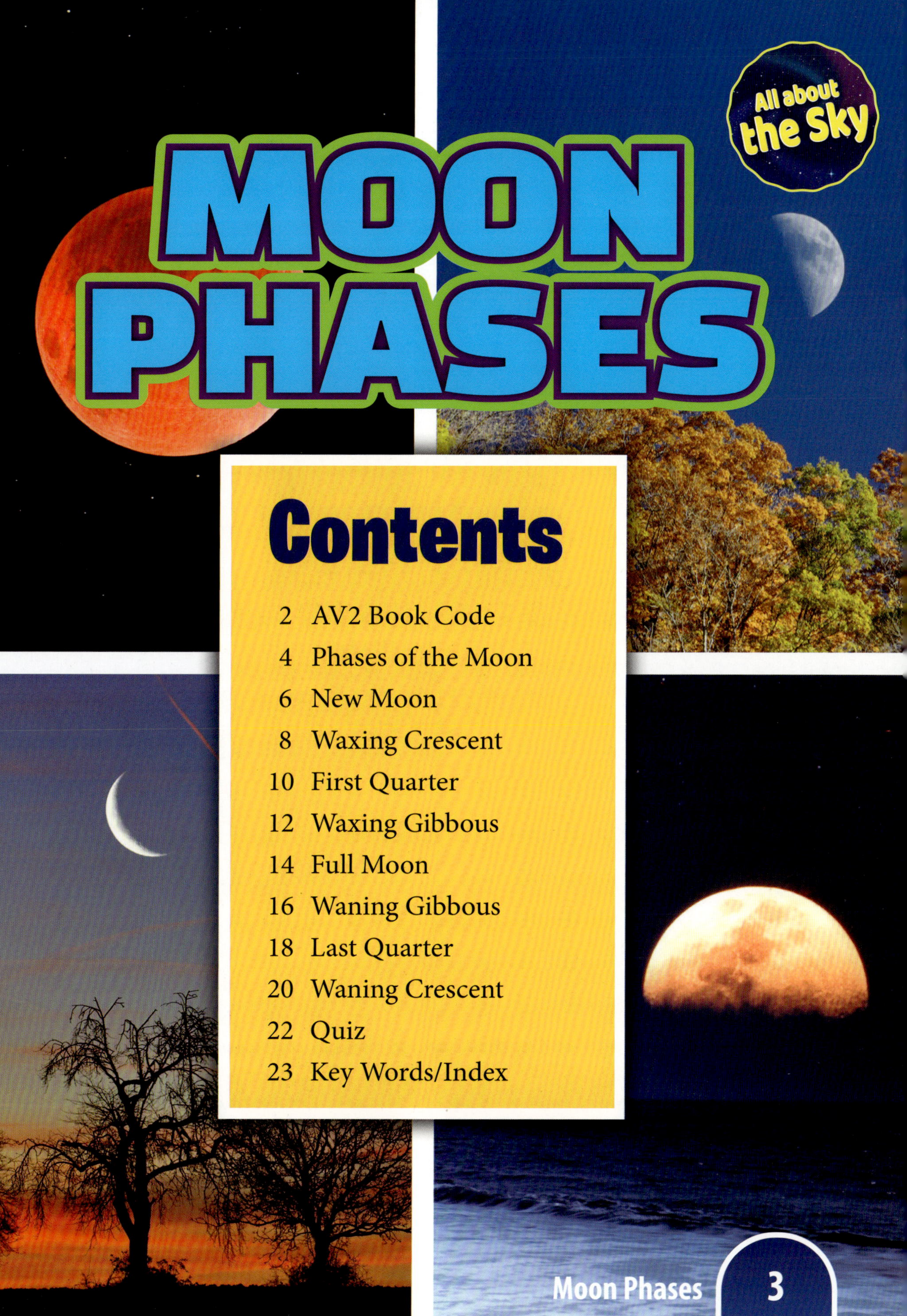

Contents

Phases of the Moon

It takes the Moon slightly less than a month to **rotate** on its **axis**. This is one day on the Moon. The Moon **revolves** around Earth in the same amount of time. This is the Moon's year. The Moon's day and year are the same length. The way the Moon moves means that the **near side** of the Moon always faces Earth.

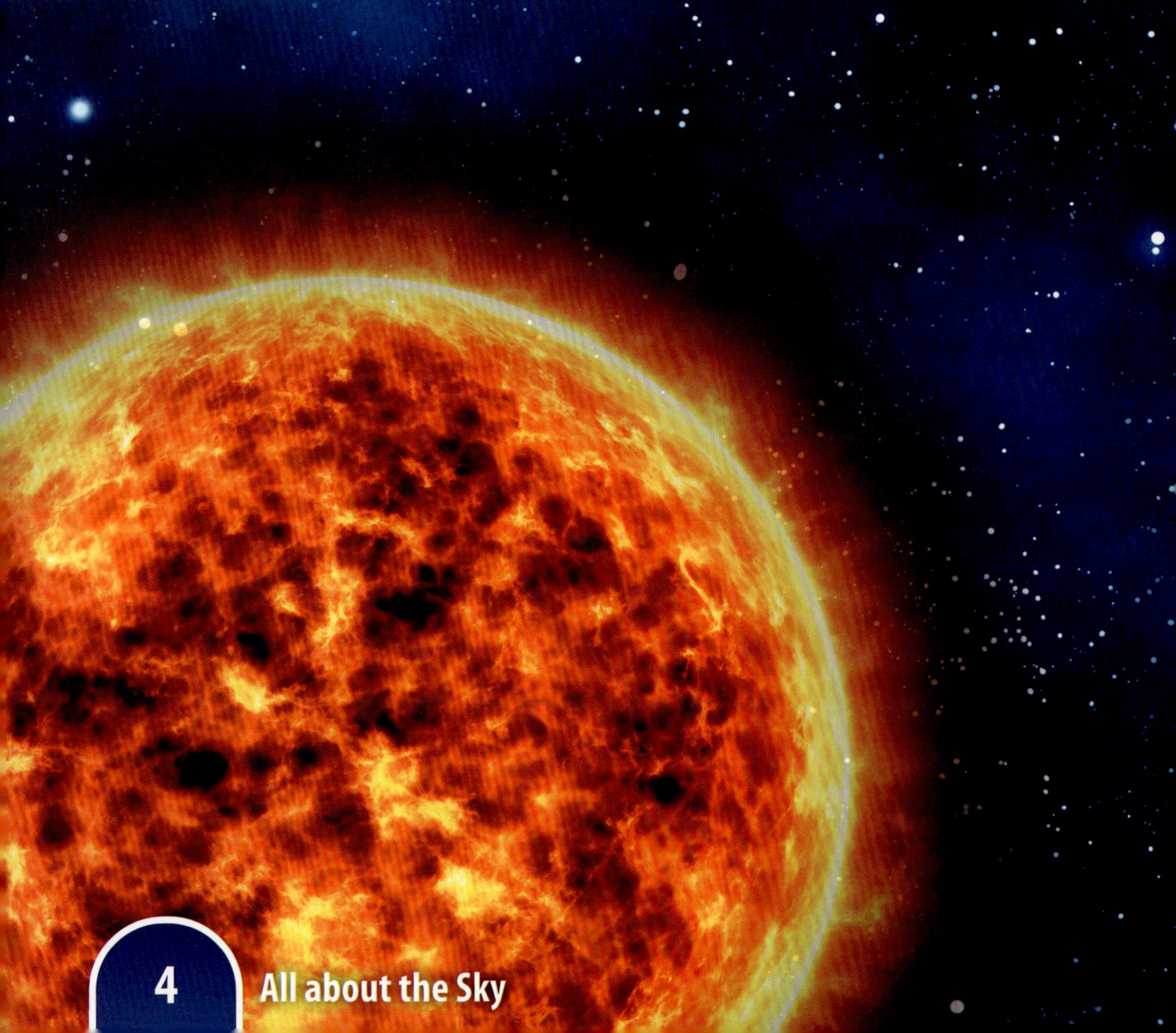

The Moon does not make its own light. Different parts of the Moon are **illuminated** by the Sun at different times of its day. These are called phases. The four **primary** phases, about a week apart, are the new moon, first quarter moon, full moon, and last quarter moon. The phases between the primary phases are called secondary phases. These are the waxing **crescent moon**, waxing **gibbous moon**, waning gibbous moon, and waning crescent moon.

New Moon

During the new moon, there is no Moon in Earth's night sky. The Moon is on the other side of the planet, where it is day. During daylight hours, the near side of the Moon is invisible because it is not illuminated. The Sun is illuminating the **far side**.

The new moon rises in the east with the Sun and sets with the Sun in the west. The Sun and Moon stay close together in the sky.

New Moon

Sometimes, the Moon crosses in front of the Sun and blocks out some or all of its light. This is known as a solar eclipse.

Waxing Crescent

As the Sun rises over the Moon's surface, the edge of the Moon is illuminated by a thin, curved sliver of sunlight. This shape is called a crescent. In the Northern **Hemisphere**, this crescent is on the right side of the Moon.

The crescent moon waxes, or grows wider, for about a week. A waxing crescent moon usually rises before noon. It cannot be seen during the day because the Sun is too bright.

Waxing Crescent Moon

A waxing crescent moon is usually visible for a while after sunset.

First Quarter

About one week after the new moon, the crescent moon waxes into the first quarter moon. The near side of the Moon is illuminated on the right. A straight line divides the Moon equally into light and dark halves.

At the first quarter moon phase, the Moon has finished a quarter of its journey. The first quarter moon rises at about noon. It is visible while the Sun is shining and sits high in the sky as the Sun sets.

First Quarter Moon

A quarter moon is often called a half moon because of its shape. However, "half moon" is not an official name.

Waxing Gibbous

The first quarter moon immediately starts to turn gibbous. The straight line separating light from dark becomes a curve that resembles a hump.

The hump waxes for another week. It gradually fills the dark half of the Moon with light. This continues until the Moon is completely illuminated. A waxing gibbous moon rises in the afternoon. It sets after midnight.

Waxing Gibbous Moon

Gibbous comes from the Latin word *gibbus*, which means "humpbacked."

Full Moon

Once the Moon reaches the full moon phase, it is completely illuminated. It looks like a bright round disc in the sky. The full moon is the brightest of the Moon's phases. It rises at sundown and sets at sunrise.

When the Moon is **aligned** with Earth and the Sun, Earth may cast its shadow on the Moon. This is called a **lunar eclipse**. A total lunar eclipse only happens during a full moon.

Full Moon

A total lunar eclipse is sometimes called a blood moon. The Moon appears to have a reddish glow.

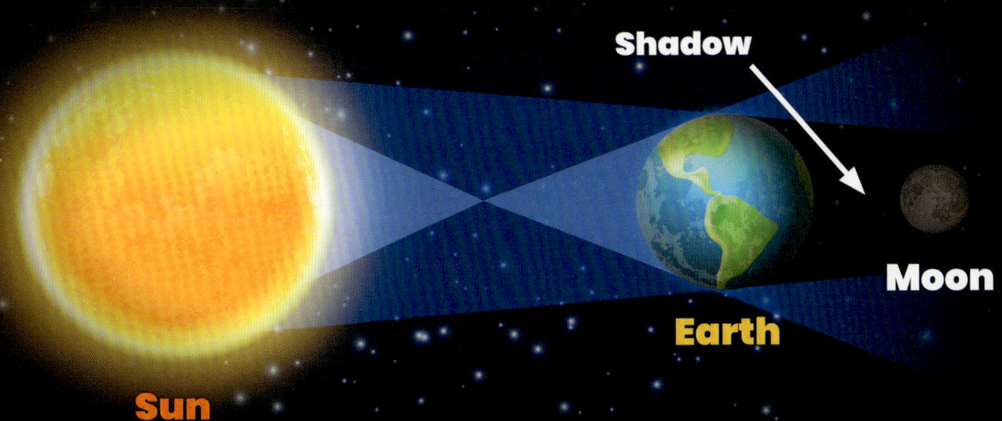

Waning Gibbous

As the Moon enters the second half of its day, its phases start to reverse. Instead of waxing, the Moon begins to wane, or get smaller. It becomes gibbous once more. The darkness moves in from the right. The gibbous hump shrinks further toward the left.

The waning gibbous moon rises after the Sun goes down. It is highest after midnight. The Moon sets after the Sun comes up.

Waning Gibbous Moon

The line that separates the Moon's light and dark halves is called the terminator.

Last Quarter

The last quarter moon is a mirror image of the first quarter moon. The light and dark sections of the Moon are again divided by a straight line at the **diameter**. This time, the left half of the Moon is illuminated.

This phase is also called the third quarter moon. Three-quarters of the Moon's **orbit** is done, along with three-quarters of its day. A last quarter moon rises at midnight. It sets around noon.

Last Quarter Moon

Earth rotates from west to east. This means that each heavenly body appears to rise in the east and set in the west.

Waning Crescent

Over the course of a week, the last quarter moon continues to wane. Eventually, only a thin sliver of light remains. This is the waning crescent moon. It looks like a waxing crescent moon, but faces the opposite direction.

A waning crescent moon rises after midnight and is at its height after sunrise. It sets after noon, invisible due to the brightness of the Sun. This completes the cycle of phases, and the next new moon approaches.

Waning Crescent Moon

The Moon's phases move from right to left in Earth's Northern Hemisphere and from left to right in the Southern Hemisphere.

Phases of the Moon

View from Northern Hemisphere

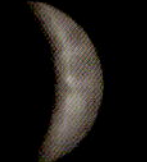 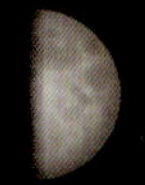 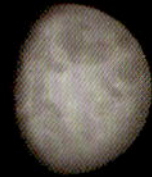  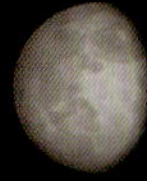 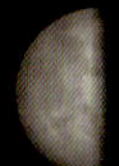

New Moon	Waxing Crescent	First Quarter	Waxing Gibbous	Full Moon	Waning Gibbous	Last Quarter	Waning Crescent

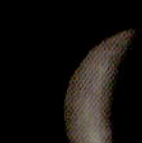 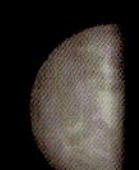 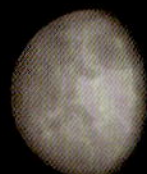 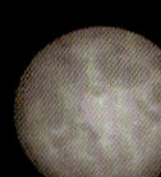 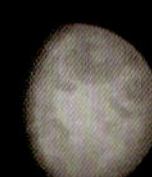 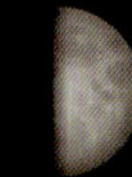 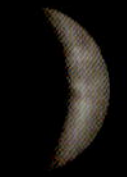

View from Southern Hemisphere

Quiz

1. Which Moon phase is the brightest?

2. What are the primary phases of the Moon?

3. What are the secondary phases of the Moon?

4. What is the last quarter moon also called?

5. What is a solar eclipse?

6. In the Northern hemisphere, on which side of the Moon is the waxing crescent?

ANSWERS

1. Full moon **2.** New moon, first quarter moon, full moon, and last quarter moon **3.** Waxing crescent moon, waxing gibbous moon, waning gibbous moon, and waning crescent moon **4.** Third quarter moon **5.** When the Moon crosses in front of the Sun and blocks out some or all of its light **6.** Right side

Key Words

aligned: arranged in a straight line
axis: an imaginary line around which a planet, moon, or star rotates
crescent moon: illuminated more than 0 percent but less than 50 percent
diameter: a straight line that divides a circle or sphere in half
far side: the half of the Moon that always faces away from Earth
gibbous moon: illuminated more than 50 percent but less than 100 percent
hemisphere: one half of Earth
illuminated: lit up
lunar eclipse: when the Moon darkens as it passes into Earth's shadow
near side: the half of the Moon that is visible from Earth
orbit: the nearly circular path that an object in space makes around a planet, moon, or star
primary: main or fundamental
revolves: moves in a nearly circular path around a planet, moon, or star
rotate: spin on an axis

Index

Get the best of both worlds.

AV2 bridges the gap between print and digital.

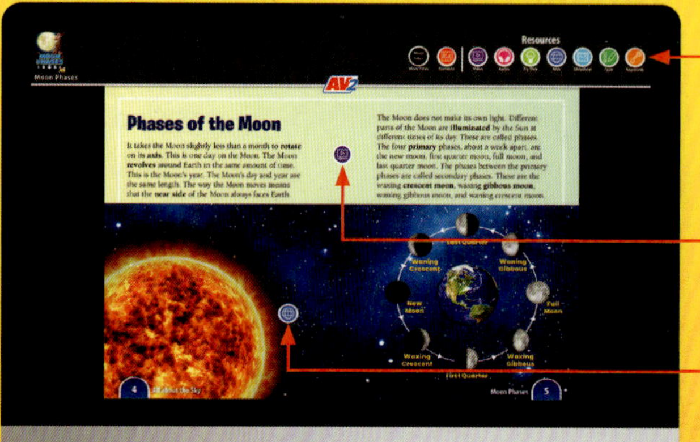

The expandable resources toolbar enables quick access to content including **videos**, **audio**, **activities**, **weblinks**, **slideshows**, **quizzes**, and **key words**.

Animated videos make static images come alive.

Resource icons on each page help readers to further **explore key concepts**.

Published by Lightbox Learning Inc.
276 5th Avenue, Suite 704 #917
New York, NY 10001
Website: www.openlightbox.com

Library of Congress Cataloging-in-Publication Data

Names: Hicton, Douglas, author.
Title: Moon phases / Douglas Hicton.
Other titles: All about the sky.
Description: New York, NY : Lightbox Learning, [2023] | Series: All about the sky | Includes index. | Audience: Grades 2-3
Identifiers: LCCN 2022011625 (print) | LCCN 2022011626 (ebook) | ISBN 9781791146160 (library binding) | ISBN 9781791146177 (paperback) | ISBN 9781791146184
Subjects: LCSH: Moon--Phases--Juvenile literature.
Classification: LCC QB588 .H53 2023 (print) | LCC QB588 (ebook) | DDC 523.3/2--dc23/eng20220611
LC record available at https://lccn.loc.gov/2022011625
LC ebook record available at https://lccn.loc.gov/2022011626

Printed in Guangzhou, China
1 2 3 4 5 6 7 8 9 0 26 25 24 23 22

072022
101121

Project Coordinator Priyanka Das
Designer Terry Paulhus

Photo Credits
Every reasonable effort has been made to trace ownership and to obtain permission to reprint copyright material. The publisher would be pleased to have any errors or omissions brought to its attention so that they may be corrected in subsequent printings. The publisher acknowledges Alamy, Getty Images, and Shutterstock as its primary image suppliers for this title.